A New Look at
Chiropractic's Basic Science

Claude Lessard, D.C.

Endorsement

The 33 Principles are the basic science of chiropractic from which our objective and its practice are derived and applied. It is the fundamental work of our profession forming the basis for our philosophy, art, and science. It is interconnected to all three. Dr. Claude Lessard has written the definitive text on the subject. The 33 Principles are the Declaration of Independence of Chiropractic which in effect causes the separation of chiropractic from the practice of medicine and every other outside-in approach. Until these principles are understood and applied, the profession will never assume its rightful place. It is a welcomed addition to the body of chiropractic literature and should be part of every chiropractor's library.

Joseph B. Strauss, D.C., F.C.S.C.

We cannot harvest what we do not sow,
and we cannot sow where we are afraid to plow.

Claude Lessard, D.C., October, 2016

I dedicate this book to Reggie Gold who was my friend and mentor, the one who continually suggested to me "to do what is right and wash my mind of all compromise."

Acknowledgments

In July of 1977, arriving in Yardley, Pa with my family to establish practice, I first met with Joseph Strauss, D.C. in his home office in Levittown. Ever since that day, Joseph has been for me an inspiration of integrity for chiropractic. The fact that he has written many books and published many articles to further chiropractic philosophy, testify to his greatness. I am very, very grateful for his constant challenge to clarify conflicts that exist in chiropractic.

This book grew from Joseph's ability to ask me hard questions. In some way or another, many of the concepts I have explored over the years focus on clarifying the authority of chiropractic and its objective, on helping people (including chiropractors) discover the firm foundation of the 33 Principles of chiropractic's basic science. Understanding WHO we can choose to be at any given moment, is a gradual evolutionary process and there are many ways we can ease and speed that process. Joseph organized my educated mind into a map to guide me along the convoluted path back to the essentials. As an objective chiropractor, I fully embrace the universal major premise of chiropractic and its 32 subsequent principles as the authority of chiropractic giving us a solid foundational platform.

Many teachers, from past and present, have shaped my understanding chiropractic philosophy, science and art. D.D. Palmer, D.C.; B.J. Palmer, D.C.; R.W. Stephenson, D.C.; Reggie Gold, D.C.; Thom Gelardi, D.C.; Joe Flesia, D.C.; Guy Riekeman, D.C.; Joe Strauss, D.C.; Joe Donofrio, D.C.; Jim Healey, D.C. and all teachers of chiropractic philosophy.

Thank you to Judy Campanale, D.C. for editing this book and helping me clarify my own thoughts and the deductive process involved in the concepts of this work.

Thank you to Tom Ruiz who guided me with the data processing analogy so necessary to this work.

Finally, thank you to my wife Sara and my children Tara, Jeremy and Sabrina who are a constant re-minder that chiropractic is truly amazing.

We shall not cease from exploration,
And the end of all our exploring
Will be to arrive where we started
And know the place for the first time.

T.S. Eliot

Preface

Chiropractic is a philosophy, a science and an art. Those three components are taken for granted in present debates on policy within the profession and in chiropractic colleges. As new information is made available to chiropractors, it is clear that the art portion of chiropractic has evolved over past decades in terms of analysis and adjusting procedures. Philosophy also evolves with new information, and in chiropractic it has advanced from being therapeutic (getting sick people well) toward non-therapeutic (location, analysis and correction of vertebral subluxation regardless of the conditions), even though it is embraced by only a small segment within the profession. A third component of chiropractic, science, is less well defined. This book defines and discusses this important third aspect of chiropractic and its impact on the profession as a whole.

Introduction

Chiropractic science is a term with positive connotations, important for example in developing the scope of practice from state to state. The term generally corresponds to anatomy, physiology, chemistry and physics in the academic world, but also includes "mathematical change measurements" and to a certain extent, research. In chiropractic discourse these distinctions are regularly overlooked. The profession's neglect of the differences corresponds to a trend in chiropractic studies. Within present historical, philosophical, sociological, and economic, studies of chiropractic there is a strong tendency to play down the difference between basic and applied sciences, or between science and technology, or explicitly to reject it as relevant to the politics or governance of chiropractic. According to Seton Hall University's Department of Science and Technology Center, it is precisely "Basic science that is concerned with the process of discovery. Basic scientists seek to discover new knowledge and information without the primary concern of how the principles they uncover might be used. Applied science takes information that already exists, and utilizes it for the solution of an existing problem. All scientific disciplines (physics, chemistry, biology, psychology, etc.) have basic and applied aspects. Basic science is more fundamental in the sense that without discovery of PRINCIPLES (emphasis mine) there is nothing to apply. Applied science relies on and could not exist without basic science."

This book will demonstrate that the 33 PRINCIPLES of chiropractic were discovered by chiropractors and are in fact, the solid foundational platform of chiropractic's basic science. It also argues that those principles are absolutes, and that if chiropractors were to study each one in depth, they would come to the realization that they are the authority on which chiropractic is based, and that they can be trusted to be a firm guide to the practice of chiropractic. This would improve the chances of developing a chiropractic profession to serve society as a whole, not only the special interests of certain groups, and perhaps could be grounds for union amongst chiropractors themselves.

It is recommended that you, the reader, watch the companion video to this book prior to reading. The video is an introduction to the book's material by the author, Dr. Claude Lessard.

A New Look at Chiropractic's Basic Science Introduction Video
https://youtu.be/kzhzmqEMJlk

Background

Science is divided in two categories: Basic science and Applied science.

Basic science is concerned with the process of discovery. Basic scientists seek to discover NEW knowledge and information without the primary concern of how the knowledge they create might be used, for example, mathematics, one plus one equals two... ALWAYS. It is absolute, duplicable and constant.

Applied science takes information that already exists and utilizes it for the solution of an existing problem. For example, chemistry will apply the basic science of mathematics to describe a molecule of water as being comprised of two atoms of hydrogen and one atom of oxygen (H_2O).

All scientific disciplines, chiropractic, physics, chemistry, biology, psychology, etc., have basic and applied aspects. Again, basic science is primary in the sense that without discovery of PRINCIPLES, there is nothing to apply. Applied science relies on and could not exist without basic science. For example, aviation uses two basic laws, the law of gravity and the law of aerodynamics (lift and thrust), and applies them to its objective which is flying. In the same way, chiropractic maintains two basic laws, the law of organization (universal intelligence) and the law of ACTIVE organization (innate intelligence). When applied, its reasoned objective is to locate, analyze and facilitate the correction of vertebral subluxations for a full expression of the innate forces within the body. Period.

Professor Alexander Spirkin, Vice-president of USSR Philosophical Society, mentioned in one of his books, *The Fundamentals of Philosophy* (1990), that:

"Science and philosophy have always learned from each other. Philosophy tirelessly draws from scientific discoveries

fresh strength, material for broad generalizations, which to the sciences it imparts the world view and methodological IMPULSES of its universal PRINCIPLES."

What's more, according to Webster's dictionary the definition of principle is: "basic truth, an idea forming the basis of something." There is therefore a necessary link between basic science and philosophy in chiropractic, where the fundamental truths of its principles are imparted a worldview, *Above Down, Inside Out* (ADIO), which leads to a specific objective.

As early chiropractors, specifically D.D. Palmer, B.J. Palmer, and R.W. Stephenson, observed universal organization, they saw a universal PRINCIPLE, and through inductive reasoning they formulated a universal major premise: "A universal intelligence is in all matter and continually gives to it all its properties and actions, thus maintaining it in existence."

It must be noted that the major premise is a UNIVERSAL PRINCIPLE and belongs to everything universal, not just chiropractic. Any field of endeavor, can appropriate this universal principle and work with it. For chiropractic's basic science, this apriori IS chiropractic's starting point. It is chiropractic's "1+1=2!" Therefore, the major premise is ABSOLUTE, DUPLICABLE AND CONSTANT!

In science, there are two ways of arriving at a conclusion: deductive reasoning and inductive reasoning. Deductive reasoning is used when a researcher works from the more general information to the more specific. Sometimes this is called an "above-down" approach because the researcher starts at the top with a very broad spectrum of information and works his/her way down to a specific conclusion. An example of deductive reasoning can be seen in the set of statements: EVERY DAY, I leave for work in my car at seven o'clock. EVERY DAY, the drive to work takes thirty minutes and I arrive at work on time. Therefore, I conclude that if I leave for work at seven o'clock today, I will be on time. The deductive statement above is a perfectly logical statement, and does rely on the initial premise being correct. The key word is EVERY DAY. Perhaps if today there is an accident on the way to work, slowing down traffic, you may end up being late. This is why any hypothesis can never be completely proved, since there is always the possibility that the initial premise could be wrong.

Inductive reasoning works the opposite way, moving from specific observations to broader generalizations and theories. This is sometimes called a "below-up" approach. The researcher begins with specific observations and measures, begins to detect patterns and regularities, formulates some tentative hypotheses to explore, and finally ends up developing some general conclusions or theories. An example of inductive reasoning can be seen in this set of statements:

Today, I left for work at seven o'clock and I arrived on time. Therefore, every day that I leave the house at seven o'clock, I will arrive at work on time.

While inductive reasoning is commonly used in science, it is not always logically valid because it is not always accurate to assume that a specific observation is correct. In the example above, perhaps "today" is a Saturday with less traffic, so if you left your house at seven o'clock on a Monday, it would take longer and you would be late for work. It is illogical to assume an entire premise just because one specific observation seems to suggest it.

Chiropractic's basic science uses mostly deductive reasoning, and some inductive reasoning, to establish its 33 Principles.

Since a principle is forming the basis of something, the 33 Principles of chiropractic's basic science form the foundational platform from which the chiropractic objective is derived. Those 33 Principles must be understood in light of that objective.

Through observation, rationalization and reasoning, an "a-priori" statement is formulated and established. Based on this assumption, using rational logic, we deduce principles that give rise to a basic science which in turn reveal the chiropractic objective. If what we assume from our observation is true, then we formulate our statement. If our "a-priori" statement is true and our reasoning is correct, we can be assured that our conclusions are correct.

The 33 Principles

As we observe the organization of the universe and the organization of its content comprised of electrons, protons and neutrons, we realize that the configuration and velocity of each particle organized into atoms is an effect of a cause. This cause stems from the fact that every effect must have a cause and that the effect depends upon the cause for its existence, and furthermore, that nature cannot produce itself. Basically, nothing can arise out of nothing.

Thus, as we observe the existence of the universe, we also observe the law of organization that maintains the existence of the universe and its physical content. Organization bespeaks intelligence. There cannot be organization without intelligence. The highly complex organization of the configurations and velocities of electrons, protons and neutrons of the universe is an effect, which connotes a universal cause that we call, universal intelligence.

So, this universal principle, that we call the major premise is:

"A universal intelligence is in all matter and continually gives to it all its properties and actions, thus maintaining it in existence."

It is the start point of chiropractic and chiropractors deduce a meaning of life (existence). from it   As chiropractic's starting point, it is chiropractic's "1+1=2!" Therefore, the major premise is ABSOLUTE, DUPLICABLE AND CONSTANT! From this place, you use reasoning, specifically deductive reasoning, to elaborate on the outcomes, meanings and manifestations of this one basic truth.

So, the "discovery" of a universal principle that we call the major premise is NEW knowledge and information, which is the concern of our basic science. Again, this universal principle, that we call the major premise, is absolute, duplicable and constant. It is APPROPRIATED as the foundational platform of Chiropractic's basic science. We can apply the information from this platform to extrapolate principle #2:

"Expression of this intelligence through matter is the chiropractic meaning of life" (which is really universal life also called existence).

As we declare the chiropractic meaning of life (as I mentioned we understand the meaning of "life" as being universal life, which is existence and comprises *all* of matter), through deductive reasoning, we can conclude that:

"Life (existence) is necessarily the union of intelligence and matter"

This is principle #3. This will be elaborated on to illustrate the process of deductive reasoning and to lay the groundwork to make the case that the 33 Principles ARE our basic science.

Having accepted that universal intelligence gives to matter all of its properties and actions, and since we know that matter depends upon an organizing intelligence … therefore there must be something that unites the two. We can conclude that existence (universal life) is a triunity having three necessary united factors, namely:

1. INTELLIGENCE (giving of properties and actions)

2. FORCE (the factor uniting intelligence and matter)

3. MATTER (being maintained in existence)

We deduce principle #4, which is:

"Life (existence) is a triunity having three necessary united factors, namely, intelligence, force and matter."

We must take note that the triune consists of two metaphysical components, (intelligence and force) and one physical component (matter).

Since matter is maintained in existence by universal intelligence;

and that if matter travels at the speed of light squared, it is transformed into energy ($E=mc^2$). It is thus reasonable to conclude that matter and energy possess all the same elements of the universe. Since according to the law of conservation of mass and energy, energy and matter are never created nor destroyed, this leads us to deduce principle #5 which states:

"In order to have 100% life (existence), there must be 100% intelligence, 100% force, 100% energy/matter."

At this time I will be using the term E/Matter in order to remind us that energy and matter are both the same physical entities of different forms.

As each of the three components of the triune are 100%, this means that they are ALWAYS whole within the universe within which EVERYTHING is COMPLETE (never created nor destroyed). Therefore we can conclude from deductive reasoning and rational logic that principle #5, which points to "the perfection of the triune," is ALWAYS perfectly whole and is stated as follow: "In order to have 100% life (existence,) there must be 100% intelligence, 100% force, and 100% matter." It really is ALWAYS so in the universe! Intelligence, force and E/Matter are always 100% within the universe since E/Matter is never created nor destroyed. In other words, the universe is always completely whole. The triune of life (existence) is perfect.

At this TIME since we introduced the PROCESS of measurement (completeness versus incompleteness) of principle #5 as being "100%" and that universal intelligence CONTINUALLY gives properties and actions to E/Matter, we deduce that:

"There is no process that does not require time,"

This is principle #6 and is called the "principle of time." It must be noted that time is accounted for within the major premise when it states that "a universal intelligence IS in all matter CONTINUALLY GIVING to it all its properties and actions, thus maintaining it in existence." Yet it is introduced after the PROCESS of completeness as principle #6, just before introducing the functions of intelligence, force and E/Matter. (It reminds us of the necessity of time for functions to occur.)

Of course, if principle #5 is ALWAYS 100%, then we deduce that the amount of intelligence for any amount of E/Matter is 100%, and is always proportional to its requirements which principle #7 states:

> *"The amount of intelligence for any given amount of E/Matter is 100%, and is always proportional to its requirements."*

E/Matter requires specific properties and actions to be maintained in existence (pri. 1), which in turn requires perfect intelligence (pri. 5) in order to provide those properties and actions of E/Matter perfectly, all the time. Example: there is 100% intelligence giving all the properties and actions (strong bonds) to the E/Matter of a beam of steel. There is also 100% intelligence giving all the properties and actions (weaker bonds) to the E/Matter of a beam of oak tree. The requirements of steel and oak are different. The intelligence is ALWAYS 100% within both.

The giving of properties and action of E/Matter to be maintained in existence by intelligence is the function of the specific configuration of their electrons, protons, neutrons and their specific velocity. This must be accomplished through specific information (force), created by intelligence, in order to maintain its form of E/Matter. Therefore, we conclude through deductive reasoning that the function of intelligence is to create force (instructive information), which is deduced as principle #8 stating that:

> *"The function of intelligence is to create force."*

Since intelligence is metaphysical, its function is also immaterial. Therefore, through deductive reasoning and rational logic we conclude that force, being created by metaphysical intelligence, is also immaterial. Force, being metaphysical, CANNOT be energy since energy and matter are comprised of the same elements of the universe and as such, are interchangeable ($E=mc^2$) and can be measured. On the other hand, force in chiropractic, is essential INFORMATION, which is strictly metaphysical, and therefore not measurable. It is INSTRUCTIVE information that is created by intelligence and is ALWAYS 100%, which is stated as principle #9:

"The amount of force created by intelligence is always 100%"
(which means perfect and COMPLETE.)

This literally means that metaphysical intelligence creates an immaterial force (instructive information) as its function. Again, we see that, within the triune, force is NOT energy, since energy is a form of physical matter ($E=mc^2$).

From intelligence, force, and E/Matter comes existence and can only occur when the second component of the triune UNITES intelligence and E/Matter. It is information (force) that the configures electrons, protons, neutrons and their velocity of E/Matter, thereby coordinating properties and actions, to maintain its form of existence (pri. 1). It is this instructive information (called force) that UNITES intelligence and E/Matter which is deduced as principle #10 stating that:

"The function of force is to unite intelligence and E/Matter."

This is precisely when the metaphysical realm interfaces with the physical realm. Therefore we logically conclude that the activity of the electrons, protons and neutrons is both metaphysical and physical which will be explained later in principle #13.

When intelligence creates instructive information (force), which unites intelligence and E/Matter, physical laws are manifested by E/Matter. These physical laws are constant and absolute; they are a manifestation of universal force which are unswerving and unadapted. Those universal forces are continually configuring the electrons, protons, and neutrons of E/Matter (properties) and their velocity (actions) UNITING intelligence and E/Matter in order to maintain E/Matter in existence. It does this regardless of the specific structure of E/Matter, which is the nature of universal forces as stated in principle #11:

"The forces of universal intelligence are manifested by physical laws; are unswerving and unadapted, and have no solicitude for the structures in which they work."

In other words, universal forces cannot create nor destroy E/

Matter as stated by Newton's law of conservation of mass and energy. It is the structure of E/Matter that can be transformed into DIFFERENT, SPECIFIC states that is the result of limitations of E/Matter. DIFFERENT, SPECIFIC configurations of the electrons, protons and neutrons (properties) of the E/Matter and their velocities (actions) will give rise to DIFFERENT, SPECIFIC structures of E/Matter, all of them having DIFFERENT, SPECIFIC limitations.

We see the above principles confirming that E/Matter is dependent on information (force) created by universal intelligence to be maintained in existence (pri. 1). Universal existence of E/Matter has unlimited possibilities within infinite probabilities of DIFFERENT configurations of electrons, protons and neutrons (properties) of E/Matter and their velocities (actions). Every specific action of E/Matter is bound by time (pri. 6), since it requires MOTION from point A to point B. Therefore, we conclude that:

"There can be interference with TRANSMISSION of universal forces."

This is principle #12. For example, shade from the leaves of a tree will interfere with the TRANSMISSION of the universal force of the rays of the sun and will prevent certain sun-needing plants from thriving under its shade. Your house will interfere with the elements of a storm passing by and will keep you safe. Interestingly, interference is not good or bad, it just is.

We finally arrive at the crux of the MATTER, which is the function of E/Matter. E/Matter is maintained in existence (pri. 1) through the instructive information that unites intelligence and E/Matter (pri. 10). Therefore, we deduce that the function of E/Matter, which is physical, is to express the organizing intelligence (force), which is metaphysical. Thus principle #13 states that:

"The function of matter is to express force" which is instructive information.

This is the interface of the metaphysical realm with the physical realm. This is exactly where the immaterial and material meet, maintaining ALL of the content of the universe in existence. This is fundamentally consistent with the chiropractic objective as will be demonstrated later.

The expression of intelligence is manifested through MOTION in E/Matter as specific properties and ACTIONS of E/Matter are derived from the specific instructive information (forces) of intelligence. Principle #14 is thereby deduced:

"Force is manifested by motion in matter; all matter has motion, therefore there is universal life in all matter."

It is worth noting that it is through the actions of E/Matter (velocity of electrons, protons and neutrons), as stated in the major premise, that the principle of time is revealed as MOTION in E/Matter (pri. 1 and 6).

Of course, without universal intelligence continually giving E/Matter all of its properties and actions, there would be no MOTION since the configuration of electrons, protons, and neutrons of E/Matter and their velocity would be absent. Therefore, E/Matter can have no motion without the coordinating information (force) created by intelligence and would cease to exist. This deduction is expressed as principle #15:

"Matter can have no motion without the application of force by intelligence."

Since principle #1 states that universal intelligence maintains ALL E/Matter in existence, including both non-living E/Matter and living E/Matter, we deduce, through rational logic, that information (force) conveys the intelligence that configures the electrons, protons, and neutrons and their velocity, of both organic (living) and inorganic (non-living) E/Matter. This is stated as principle #16:

"Universal intelligence gives force to both organic and inorganic matter."

In other words, universal intelligence IS in ALL E/Matter (living and non-living) as per principle #1. We must note that living E/Matter and non-living E/Matter are observed aspects of different states of organization and are introduced, within principle 16, through inductive reasoning.

The first 16 principles point to a universal intelligence CONTINUALLY giving to ALL E/Matter ALL of its properties and actions in order to maintain ALL E/Matter in existence (the major premise pri. 1). Therefore, universal intelligence is the CAUSE of all E/Matter to be MAINTAINED in existence. In other words, universal intelligence is the CAUSE of the MAINTENANCE of E/Matter in existence, which is the EFFECT. From this cause and effect, we deduce principle #17 stating that:

"Every effect has a cause and every cause has effects."

In principle #16 we saw that universal intelligence gives instructive information (force) to both living E/Matter and non-living E/Matter in order to maintain ALL E/Matter in existence. The distinction between living E/Matter and non-/living E/Matter is evidenced through specific characteristics arising from active organization, called the signs of life, and consists of principle #18 which states that:

"The signs of life are the evidence of the intelligence of life."

It is through inductive reasoning that those signs of life are observed and identified as assimilation, excretion, adaptability, growth and reproduction. All of these signs must be present in order for a thing to be living.

It is reasonable to deduce principle #19 from the previous principles since ALL E/Matter is maintained in existence through specific instructive information (force) which configures the specific electrons, protons, and neutrons of E/Matter and their velocity in an organized manner, as to interrelate with ALL universal E/Matter. Therefore, through rational logic, we deduce principle #19 which states that:

"The material of the body of a "living thing" is organized E/ Matter."

The organized E/Matter of a thing that is living has a different level of complexity of structural organization and has different motion than a non-living thing. It is the evidence of the signs of life that points to the intelligence of life, which is shown in principle #18. These signs of life require a specific level of instructive information CAUSED by a specific intelligence capable of ADAPTING universal instructive information (force) and E/Matter for it to live. This specific intelligence is called: innate intelligence. Therefore, principle #20 states that:

"A "living thing" has an inborn intelligence within its body, called innate intelligence."

Innate intelligence CAUSES a level of structural organization of living E/Matter different from non-living E/Matter and is the evidence of the signs of life of a living thing, as per principle #18. This level of structural organization is an action of innate intelligence and maintains the E/Matter of the living thing alive. Thus, we deduce principle #21 which states:

"The mission of innate intelligence is to maintain the material of the body of a 'living thing' in active organization" which means to maintain life in the body of a living thing.

Intelligence is metaphysical, is ALWAYS 100% perfect and complete in ALL E/Matter to maintain its existence, including "living things", according to their specific level of complexity of structural organization (pri. 7). This is principle #22:

"There is 100% of innate intelligence in every 'living thing,' the requisite amount, proportional to its organization."

This means that ALL of the needs of living E/Matter are ALWAYS perfectly met, within the limits of adaptation of E/Matter. In other words, there is an innate awareness for every innate need.

ACTIVE organization denotes a different level of organization. One that evidences the signs of life as opposed to one that does not (your body as compared to a stone). The law of organization, universal intelligence, gives ALL properties and actions to ALL E/Matter and maintains it in existence (pri. 1). The law of ACTIVE organization maintains the E/Matter of a "living thing" alive (pri. 21), demonstrating the signs of life (pri. 18). For E/Matter to evidence the signs of life, there has to be coordination of activities of the components of the living E/Matter. Therefore principle #23 states that:

"The function of innate intelligence is to adapt universal forces and E/Matter for use in the body, so that all parts of the body will have co-ordinated action for mutual benefit."

ACTIVE organization deals with ALL "living things" and is ALWAYS 100% in any "living thing" proportional to its structural organization (pri. 22). Therefore, since innate intelligence is ALWAYS perfect (100%), principle #24 states:

"Innate intelligence will adapt forces and E/Matter for the body as long as it can do so without breaking a universal law,"

This is principle #24. We must always remember that "ALL parts of the body" signify multi levels of organization; the innate intelligence of the body, the innate intelligence of systems (cardiovascular, respiratory, endocrine, etc.), the innate intelligence of organs (liver, heart, kidneys, etc.), and the innate intelligence of cells. For this reason, there are multi levels of limitations of "living things." For example, a polar bear in Antarctica will sustain subfreezing temperature, yet a panther of East Africa will not. On the other hand, a panther in East Africa will sustain scorching temperature, yet a polar bear of Antarctica will not. The same can be said for an adult that can eat a complete cooked meal and get nutrition from it, while a two-day old infant could not.

In other words, innate intelligence is constrained by the limitations of E/Matter. We must understand that, within principle #24 is assumed the limitation of time. An example would be the digestion of a meal. It requires time to properly digest a meal. If you introduce food before the previous digestion is completed, you may wind up vomiting, not necessarily because of the limit of your E/Matter, but because of the amount of time that your E/Matter requires to digest a meal. In other words, innate intelligence is subject to the limitation of E/Matter and time.

Innate intelligence being 100% perfect and complete in ALL

"living things" adapts universal forces and E/Matter for use in the body, WITHOUT BREAKING A UNIVERSAL LAW, so that all parts of the body will have coordinated action for mutual benefit. This discloses the character of innate forces as being continually consistent in upholding the integrity of the structure in which they act. In other words, the instructive information from innate intelligence toward the structural E/Matter to maintain it in active organization is always congruent with keeping the "living thing" alive. Therefore principle #25 states that:

"The forces of innate intelligence never injure or destroy the structure in which they work."

Now, if we were to compare universal forces which maintain E/Matter in existence (pri. 1) with innate forces which maintain the matter of the body of a "living thing" in active organization (pri. 21), we would deduce that the flow of life depends on innate intelligence adapting the configuration of the electrons, protons and neutrons of E/Matter and their velocity. This adaptation, from innate intelligence, is maintaining the material of the body alive, which is the evidence of the signs of life. On the other hand, universal intelligence continually creates universal forces for the purpose of maintaining ALL E/Matter in existence. This includes the material of the body, in which universal forces configure the electrons, protons and neutrons of E/Matter, and their velocity toward their most stable state, which is the atomic state. In other words, in order to carry on the universal cycle of life:

"Universal forces are 'destructive,' and innate forces constructive, as regards structural E/Matter"

This is principle #26. This principle is also congruent with the law of conservation of E/Matter in so far as E/Matter can never be DESTROYED nor created. Therefore, "destructive" in the context of principle #26 implies DECONSTRUCTION of structural "living" E/Matter of the body toward its most stable state, which is at its atomic level.

Since innate intelligence is 100% perfect and complete (pri. 22), adapting universal forces and E/Matter for the body (pri. 23) without breaking a universal law (pri. 24) in order to maintain the material of the body alive (pri. 21), we deduce that the law of active organization is ALWAYS normal. 100% perfection and complete is the NORM regarding the law of active organization. Therefore principle #27 states that:

"Innate intelligence is always normal and its function is always normal."

At this point, it is necessary to discuss, some fundamental understanding of metaphysical realities. First of all, a universal intelligence is in ALL E/Matter (pri. 1). Universal intelligence is the LAW of existence, also called the LAW of organization. Without it, E/Matter would not be maintained in existence, and therefore, E/Matter would not exist. Universal intelligence creates ALL forces ALWAYS. It is the source of unlimited forces that maintains E/Matter in existence.

Now, as regards to living organisms, a "living thing" has an inborn innate intelligence (pri. 20). Innate intelligence is the LAW of "living things," also called the LAW of active organization and is the source of unlimited, adapted universal forces into constructive forces for ALL the tissue cells. We must remember that innate intelligence has an unlimited supply of universal forces at its disposal. Innate intelligence is a part OF universal intelligence. Innate intelligence is a part FROM universal intelligence also, in so far as it is localized within the body of a "living thing."

Basically, the LAW of active organization is a localized portion of the LAW of organization, and as such, is "under" the authority, so to speak, of universal intelligence. As an example, Newton's third law of equal and opposite reaction is "under" the authority of his second law of acceleration. Therefore, innate intelligence, which is metaphysical, is "under" the authority of universal intelligence, also metaphysical, for the maintaining of E/Matter in existence. Universal intelligence is EVERYWHERE in the universe and innate intelligence is EVERYWHERE within the "living thing." There are NO connectors between innate intelligence and universal intelligence since universal intelligence is EVERYWHERE.

In order for the constructive forces of innate intelligence to coordinate the action of all the parts of the body for mutual benefit to occur (pri. 23), they must be first be assembled before being centralized and transmitted. These constructive forces are called mental impulses. Innate intelligence has an immaterial "headquarters" within which to operate. Here, the innate brain is introduced as a system of operations, which is a metaphysical concept, comprised of coded and programmed information. There innate intelligence assembles innate forces that will be transmitted to the physical brain for centralization and distribution. The reason why the innate brain is metaphysical, is because the innate brain is the seat of innate intelligence which is EVERYWHERE within the body. Therefore, the "headquarters" of innate intelligence is the innate brain and must be wherever innate intelligence is, which is EVERYWHERE within the living body!

At the precise moment that some innate forces are coded and programmed for coordination of action of all the parts of the body (pri. 23), they are now mental impulses, assembled within the innate brain.

We saw above, that principle #27 states that the function of innate intelligence is ALWAYS normal and that one of the purposes of that function is to have coordination of action of all the parts of the body for mutual benefit (pri. 23). In order to have coordination of action, there must be interconnectivity between ALL the parts in order to manifest mutual benefit of ALL the parts. In other words, as innate intelligence adapts universal forces and E/Matter for coordination of action (pri. 23), the congruency of instructive information necessary to coordinate all the parts of the body needs intelligent interconnectivity.

In order to introduce our next principle #28, I will use computer data processing as analogies to demonstrate the transmission of innate forces within the living animal body. This will parallel the "Normal Complete Cycle" showing intellectuality perpetuated in cycles (Chiropractic Textbook, R. W. Stephenson, p.336 art. 398.)

NORMAL COMPLETE CYCLE, R.W. Stephenson, P.11, Fig.5

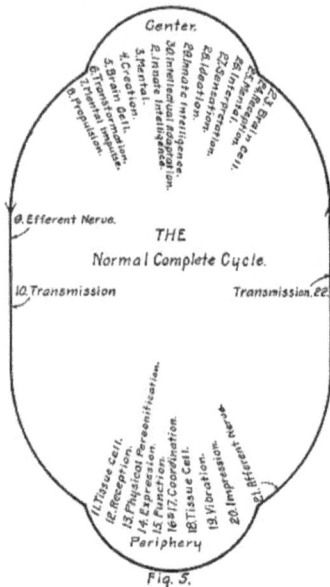

CHIROPRACTIC TEXTBOOK
BY RALPH W. STEPHENSON, D.C., PH.C.
Illustrations by the Author

1 - Universal intelligence.
2 - Innate intelligence.
3 - Mental.
4 - Creation.
5 - Brain cell.
6 - Transformation.
7 - Mental impulse.
8 - Propulsion.
9 - Efferent nerve.
10 - Transmission.
11 - Tissue cell.
12 - Reception.
13 - Physical Personification
14 - Expression
15 - Function.
16 - Coordination.

31- Universal intelligence
30 - Intellectual adaptation.
29 - Innate intelligence.
28 - Ideation.
27 - Sensation.
26 - Interpretation.
25 - Mental.
24 - Reception.
23 - Brain cell.
22 - Transmission.
21 - Afferent nerve.
20 - Impression.
19 - Vibration
18 - Tissue cell.
17 - Coordination.

From what scientists have learned over the years, from data processing, "fields programmable gate arrays" consist of a large number of "logic block" programs that can be configured and reconfigured, individually, to do a wide range of tasks. One logic block may do arithmetic, another signals processing, and yet another look things up in a table. The computation of the whole is a function of how the individual parts are configured. According to E-XILINX, "field programmable gates arrays" can be reprogrammed to the desired functionality requirements AFTER manufacturing.

We now know that the language of neurons and synapses are connected to bodily functions. It appears that neurons are akin to computer hardware and bodily functions are akin to the actions that a computer performs, which means that computation of information is probably uniting the two. If the heart is a biological pump, and the nose is a biological filter, then the physical brain with its nerves is a biological computer (i.e. a biological central processing unit), basically a sophisticated system of computing and transmitting information for coordination of actions (pri. 23).

It is difficult to see the physical brain and its nerves without preconceptions. Yet, making a significant impact requires us to close our eyes, and open them back up as if we are seeing the nerve system for the very first time. To see the physical brain, without those rules that lock us in, is to see something NEW. Therefore, to observe that the nerve system IS the system used by innate intelligence to compute and transmit its information in animal bodies is to delve into the area WHERE there can be interference with transmission of forces (pri. 12).

In order to explain this NEW observation, I will be using parallel logic to facilitate our understanding of our induction. First of all, let me state that the maxim, "we cannot give what we do not have" is true. Differently stated is the fact that whatever we "create/invent" which is a re-organization of E/Matter into EXTERNAL inventions, is ALWAYS a reflection of INTERNAL, already existing organization of structural E/Matter (pri. 21, 23 and 26). That is WHY we could manifest the many EXTERNAL inventions that we live with in the world. Examples abound, we invented the pump (heart), bridges (ligaments), filter (nose), artificial breathing machine (lungs), camera (eye), telephone (ear), irrigation (circulatory system), computer (brain), etc. Just like our major premise, it is obvious that universal intelligence and innate intelligence are the start point of our observation (Step 1 and 2 of the "Normal Complete Cycle.")

Like a computer, a control center forming the operating system is used by the innate intelligence of the body in order to coordinate the activities of all the parts of the body. As we mentioned before, this operating system is called the innate brain and it is metaphysical. We also know that the innate brain is used by innate intelligence to assemble mental impulses and that it is located wherever innate intelligence is.

Since we know from principle #6 that there is no process that does not require time, coordination of action of all the parts of the body is a process requiring time. Coordination of action also requires the INSTRUCTIVE information of innate intelligence ("Mental," step 3) ... from a physical centralization center to the parts of the body at the periphery. It needs centralization and distribution from center to periphery. Time is paramount, for physical living E/Matter, to process metaphysical data (innate forces) and to express that data (pri.6, 13 and 21). It requires time for living E/Matter to compute instructive information (pri. 23) without breaking a universal law (pri.24).

The direct contact between the expression of metaphysical data by physical E/Matter (pri. 13) and its physical motion (pri. 14) must interface with each other. This interface is the union of metaphysical intelligence and physical E/Matter expressing instructive information created by universal intelligence (pri. 10), adapted (pri. 23) and assembled by innate intelligence within the innate brain ("Creation," step 4). From its headquarters, called the innate brain, innate intelligence centralizes those mental impulses within the physical brain ("Brain cell," step 5) and TRANSMITS them through neurological matter to ALL the parts of the body for coordination of action (pri. 23).

Since it requires time to transmit data ("Transmission," step 6) to be processed by living E/Matter, the distance between different parts must be interconnected with each other through a specific central processing unit (CPU) in order to harmonize their specific function into coordinated activity. A (CPU) is the electronic circuitry within a computer that carries out the instruction (instructive information) of a computer program (specifically coded mental impulse) by performing the basic logical control and input/output (I/O) operations specified by the instructions. This centralization process, within the animal body is physical transmitting matter, which consists of a specialized system, basically micro-controllers that are comprised of specialized cells, which are microprocessors that can transmit instructive information in the form of a mental impulse through conductivity ("Mental Impulse," step 7). This specialized system is necessary to first centralize the already assembled metaphysical mental impulses, and then to propel their instructive information ("Propulsion," step 8) to the different parts of the body.

Of the many tissues comprising the animal body we observe, through inductive reasoning, that the central nervous system with its brain, spinal cord and nerves, comprise the hardware network used by the innate intelligence of the body to communicate its instructive information to all of the different parts of the animal body. From the adaptation of the E/Matter of the central nervous system by the innate intelligence of the body (pri. 23), innate forces are created to form neurons into specific applications. These neurons are coded into a multitude of data processing programs (microprocessors) interfacing with each other. The innate intelligence of the body uses the innate brain, which is the operating system, to centralize its already assembled mental impulses within the physical brain (CPU).

From the physical brain, innate intelligence uses a network of nerves as the hardware ("Efferent Nerve," step 9) to distribute the data by transmitting the mental impulse ("Transmission," step 10), which is the instructive information, through conductive energy/matter, to the parts of the body at the periphery ("Tissue Cell," step 11 and "Reception," step 12) (pri. 23). These parts in turn contain specific analytical device drivers, which are peripheral integrated circuits, like computer chips, that are used by innate intelligence to decode the instructive information, at the periphery for the coordination of actions of the parts of the body. Those device drivers are within the innate brain and are metaphysical. Stephenson's calls this decoding activity (step 13 of the universal diagram, "Physical Personification".) The interface of the metaphysical and physical occurs at the union of intelligence with E/Matter through force (pri. 10). All this is so extremely complex that the interface itself is separate and distinct (pri. 4). In other words, the interface is a part OF and a part FROM intelligence, force and E/Matter.

Suffice it to say that a highly specialized system is used by innate intelligence for the purpose of coordination of action of all the parts of the body (pri. 23). The central nervous system is the transmitting network, which is used by the innate intelligence of the body to coordinate the action of all the parts of the body for mutual benefit ("Expression," step 14) (pri. 23)... INCLUDING the central nervous system itself. It must be remembered that the central nervous system also needs coordinated action. The central nervous system is comprised of living E/Matter, called neurons, continually needing to be adapted by innate intelligence in order to perform its specific function along with all other tissue cells ("Function," step 15). Coordination of activities requires a feedback mechanism that will inform the physical brain of the response of the tissue cell. There is coordination within the tissue cell that will ultimately engender coordination of ALL the parts of the living organism ("Coordination", steps 16&17). The coordination of activities of the tissue cell will in turn express its motion ("Tissue cell," step 18) again using analytical device drivers, innate intelligence will code NEW instructive information to be transmitted back to the central processing unit, the physical brain ("Vibration," step 19). Then the NEW coded instructive information, within the innate brain, uses specific markers and drivers that will imprint the transmitting matter going back to the physical brain ("Impression," step 20). The transmitting matter carrying information from the tissue cell ("Afferent nerves," step 21) requires a specific impression to transmit the feedback ("Transmission," step 22) to the brain cell, ("Brain Cell," step 23).

Once the physical brain receives this NEW coded feedback information from the tissue cell ("Reception," step 24), innate intelligence will then use this information ("Mental," step 25) to interpret its content ("Interpretation," step 26). At this level, a series of integrated circuits within the physical brain will literally sense the feedback message of the tissue cell ("Sensation," step 27) decoding the information received from the tissue cell ("Ideation," step 28). This decoded information is automatically formulated by innate intelligence ("Innate Intelligence," step 29) for the purpose of adapting universal forces and coding them to answer the specific need of the tissue cell ("Intellectual Adaptation," step 30). The adaptation of those universal forces to fulfill the needs of the tissue cell, bring us back to universal intelligence ("Universal Intelligence," step 31), finalizing the

NORMAL COMPLETE CYCLE.

It must be noted that steps 16 through 31 are the same steps in reverse, using similar components with the afferent nerves.

Finally we arrived at principle #28 which states that:

"The forces of innate intelligence operate through or over the nervous system in animal bodies."

Now that we have used the analogy of computer data processing, it is important at this time to condense the Normal Complete Cycle.

SUMMARY OF THE NORMAL COMPLETE CYCLE, (R. W. Stephenson, p. 63, Art. 101)

"The Normal Complete Cycle is the story of what happens between cause and effect and effect and cause. The list of 31 steps is the conventional outline of the story."

The Story
(Reproduced and adapted from chiropractic's basic science.)

Universal intelligence, which is the law of organization, is in ALL E/Matter and continually gives to it all the properties and actions through the configuration of electrons, protons, and neutrons of E/Matter.

The expression of this intelligence, through E/Matter is the chiropractic meaning of existence, which is universal life; therefore the maintaining of the existence of E/Matter, is necessarily the union of intelligence and E/Matter.

Force (instructive information) unites intelligence and E/Matter. Universal intelligence (law of organization) provides instructive information (force) to both organic and inorganic E/Matter. That instructive information (force) with which the law of organization (universal intelligence) instructs the structure of organic E/Matter as a higher order of its manifested existence is called, INNATE INTELLIGENCE. This is the law of active organization.

The mission of innate intelligence (law of active organization) is to maintain the living E/Matter of an organic unit alive. It does this by adapting the forces (instructive information) of universal intelligence (law of organization). As physical laws are unswerving and un-adapted and have no solicitude for structural E/Matter; they can be used in the body so that all parts of the body will have coordinated action, thus every part has mutual benefit (pretty much like the law of gravity which is adapted by the aerodynamic design of an aircraft, by using the law of lift, the law of drag and the law of power, so it can be used for flying from point A to point B.)

This work of innate intelligence (law of active organization) is entirely MENTAL (metaphysical) and is always perfectly normal. For this reason, the instructive information (forces) of innate intelligence (law of active organization) never injures or destroys body tissues. The instructive information (forces) of innate intelligence is metaphysical (mental), and superior to physical forces (universal instructive information), because it controls, (adapts) physical forces.

This assembling of universal forces is called CREATION, the interface uniting intelligence and E/Matter, occurring within the innate brain, and having a definite form and purpose. The headquarters of innate intelligence (law of active organization), in the living body, is the innate brain (metaphysical) and uses the definite physical brain in the form of the unit called BRAIN CELL. From the brain cell as a unit, innate intelligence (law of active organization) controls a unit of E/Matter. In the brain cell, metaphysical instructive information interfaces with physical E/Matter through which innate intelligence (law of active organization) TRANSFORMS the metaphysical instructive information (mental force) into a definite unit, for a given tissue cell, for a given moment.

This specific force (instructive information), when transformed, is called a MENTAL IMPULSE. It bears specifically coded instructions, providing logical control input/output operations, setting E/Matter into action.

This specialized system is necessary to first, centralize the already assembled metaphysical mental impulses, and to propel their instructive information ("Propulsion," step 8) to the different parts of the body. The departure of the mental impulse from the brain cell is called PROPULSION.

For the mental impulse to be transmitted, from brain cell to tissue cell, requires effort. The instructive information (forces) of innate intelligence operates through or over the nerve system. That which has efferent direction (source to periphery) and which conducts the mental impulse is called EFFERENT NERVE. Since the mental impulse is transmitted via a physical medium in nature, and since physical energies (E/Matter) can suffer interruption in their transmission, in a like manner, the instructive information (forces) of the law of active organization (innate intelligence) can suffer interference with transmission. This is the basis for the existence of chiropractic. Some theories contend that the mental impulse is half a physical force and half metaphysical instructive information and therefore is subject to the same laws as any other physical force, but it should always be remembered though that these physical energies (E/Matter) are in the ADAPTED FORM of mental impulse (if this theory is used) and therefore not injurious to tissue as electricity would be.

The conveyance of the mental impulse over the efferent nerve is TRANSMISSION. Over this route of specialized neuronal circuitry, having specialized atomic elements with particular configuration and velocities, the mental impulse travels to TISSUE CELL where it is RECEIVED, whereupon the computed decoding (mental conception of innate intelligence (law of active organization), as to what the cell should be or how it should act, comes to pass. That which was only metaphysical (mental), now becomes manifested. It shows by its very character that intelligence planned the form or the action, and this evidence of intelligence is called EXPRESSION; meaning the coming forth through E/Matter; the showing of intelligence.

Things, which show this are said to be alive and such expression is called LIFE. The character of this action is determined by the character of the tool used by innate intelligence to express itself, therefore the purpose or the action of this tool, which is the tissue cell, is FUNCTION. The function of E/Matter is to express instructive information (force).

In the tissue cell, which is a specific kind of E/Matter, the instructive information (forces) of innate intelligence is expressed in a specific manner by a specific instrument (device driver) built for the particular kind of expression. The prompt and correct action of that tissue cell, being actuated by the specific instructive information (forces) of the law of active organization (innate intelligence) in harmony with all other cells is called COORDINATION. In this we see the working of the law of cause and effect, and that every process requires time.

In order to perform its function, the tissue cell has motion, both molecular and as a whole cell. This movement is called VIBRATION and codes the instructive information of innate intelligence into the response or need of the tissue cell. These vibrations give off signals, which are impressed upon the afferent nerve as a form of specific code called IMPRESSION. These impressions are transmitted over the AFFERENT NERVE. This TRANSMISSION is similar to transmission in the efferent half of the cycle, for the codes are similar. When it reaches the afferent BRAIN CELL, it is RECEIVED much in the same manner as the tissue cell receives, for the physical brain cell is a tissue cell after all.

When this instructive information has reached the brain cell it is immediately decoded by the law of active organization (innate intelligence) in order to enter the metaphysical realm; then the instructive information undergoes MENTAL INTERPRETATION, which is metaphysical. The product of this act of interpretation by innate intelligence is a SENSATION, which is a computing of instructive information. When innate intelligence has a number of computations, the resulting data is processed and the condition or need of the cell is put together by innate intelligence and this is named IDEATION, which is a clear representation of the state of the tissue cell. Ideation can be the possession of nothing but intelligence. The intelligence in the body, of course, is INNATE INTELLIGENCE. When the law of active organization deals with the state of the tissue cell, instructive information is planned to make it adapt to its environmental conditions and the metaphysical process of doing this is INTELLECTUAL ADAPTATION, which is a computation of instructive information. The great source of supply from which innate intelligence draws its information (forces) is UNIVERSAL INTELLIGENCE.

Continuing our study of chiropractic's basic science, we see that innate forces are adapted universal forces by innate intelligence into specially coded instructive information, which must be TRANSMITTED to all the parts of the body for coordination of action (pri. 23.) From our previous principle, (pri. 28,) we saw that the nerve system in animal bodies is a highly specialized network used by innate intelligence to perform the task of interconnecting all the parts of the body in order to TRANSMIT the instructive information (innate forces) to the parts.

Therefore, for innate forces to travel through or over the transmitting matter of the nerve system, which involves time and distance, innate forces must be a combination of metaphysical and physical. Hence, the function of innate intelligence is to adapt universal forces and E/Matter, (in this case, the transmitting matter) (pri. 23.) For this reason, the code of the innate force is called mental (which means that it is metaphysical) and impulse (which means that it is physical).

According to the Meriam/Webster dictionary, the word impulse means, "A small amount of energy that moves from one area to another" and the word mentality means: "Intelligence." Therefore the metaphysical coded instructive information is carried by a small amount of physical energy and transmitted to its destination for coordination of action. This physical energy results from a universal force, created by universal intelligence, that unites universal intelligence to the E/Matter of the body in order to maintain it in existence (pri. 1 and 10). We see that there is a relationship between structure and function, that if there is a change in structure, there will be a change in function. Since there can be interference with the TRANSMISSION of universal forces (pri. 12), we conclude through rational logic and deductive reasoning, that our next principle is:

"There can be interference with the TRANSMISSION of innate forces" which is principle #29.

Innate forces as we saw are adapted universal forces by innate intelligence into specific codes of instructive information that are transmitted to all the parts of the body for coordination of action (pri. 23). When there is interference with the transmission of innate forces (the mental impulse, which is constructive) en route to structural E/

Matter (pri. 26), the specific code of instructive information is lost and reverts back to simply a nerve impulse, which is a deconstructive universal force (pri. 26). The net result of the interference with the TRANSMISSION of innate forces is that the specifically coded instructive information will NOT reach its intended destination for coordination of action of all the parts of the body. This means that the intended specific message (innate force), whose purpose is coordination of action (pri. 23), will NOT be received by the parts of the body, resulting in a lack of coordination of action of the parts of the body. Therefore, principle #30 states:

> *"Interference with the TRANSMISSION of innate forces causes incoordination of DIS-EASE."*

Now that we know that innate forces operate through or over the nerve system in animal bodies (pri. 28), we deduce that the interference with TRANSMISSION of innate forces takes place within the TRANSMITTING matter, between the control center which is the physical brain and the intended recipient physical tissue cell of the body. I must STRONGLY emphasize that this interference in transmission is between brain cell and tissue cell, in other words, between E/Matter and E/Matter, NOT between intelligence and E/Matter. We also know that vertebrate animal bodies possess a spinal column, comprised of juxtaposed articulating vertebrae to protect the spinal cord, which is the major link, along with its adjacent nerves, between the brain and the parts of the body. Vertebrae are made of bones to protect this major link of data communication, for coordination of activities, within the vertebrate animal bodies. These hard bony vertebrae interrelate through a series of articulations. It is when these articulations DO NOT properly interrelate that the interference with TRANSMISSION of innate forces occurs within the transmitting matter, namely the nervous system.

It must be noted that the interference of innate forces happens WITHIN E/Matter between brain cell and tissue cell. It is the lack of proper juxtaposition of the hard bony vertebrae of the spinal column that causes a CHANGE in the physical TRANSMITTING matter of the nerve system, interfering with the TRANSMISSION of mental impulses. In other words, the CHANGE of the configuration of electrons, protons and neutrons of the TRANSMITTING matter prevents EASE of TRANSMISSION of mental impulses. At this precise moment, the TRANSMITTING matter is in a state of DIS-EASE and interferes with the TRANSMISSION of innate forces, which are coded mental impulses with intelligent direction. The lack of EASE of the TRANSMITTING matter reverts the mental impulse, which is an innate force that is constructive, to a nerve impulse, which is a universal force (deconstructive to structural E/Matter [pri. 26]). When a part of the body receives a nerve impulse instead of a mental impulse, that part will have incoordination of action. It is for this reason that principle #30 reads: "incoordination OF dis-ease." It is the physical CHANGE in the TRANSMITTING matter (nerve) that prevents EASE of TRANSMISSION of the mental impulse, leading to incoordination of action of the parts of the body. Therefore we deduce principle #31, which states:

"Interference with transmission in the body is always directly or indirectly due to subluxation in the spinal column."

It must be noted that principle #31 was evidenced in 1973 by two research studies from the University of Colorado in Boulder. In one, Dr. Chung-Ha Suh, Ph.D., developed a complex computer model of the cervical spine, which furthered a greater understanding of the mechanics of vertebral articulations and their relationship to the chiropractic adjustment. The second study involved the effect of compression on spinal nerve roots. Seth Sharpless, Ph.D., and Marvin Luttges, Ph.D., and their colleagues demonstrated that a minute amount of pressure on a nerve root, (10mm Hg), caused a 50% decrease in transmission of nerve impulse traveling down that nerve root.

The function of innate intelligence is to adapt universal forces and E/Matter for use in the body (pri. 23). "Use in the body," means primarily to keep the body alive (pri. 21), with or without interference to innate forces within the limits of adaptation (pri. 24). I mention this because it is possible for the E/Matter of the "living body" to have interference with innate forces and remain alive. Many people have vertebral subluxations uncorrected all their life and live to a very old age.

The other aspect of the function of innate intelligence is that, in the absence of interference with innate forces, it also means that all parts of the body will have coordination of action for mutual benefit (pri. 23). Coordination of action requires harmony between all the parts of the body and this harmony becomes the necessary factor for the parts of the body to fulfill their intended function and objectives. Therefore, principle #32 states:

"Coordination is the principle of harmonious action of all the parts of an organism, in fulfilling their offices and purposes."

Function is the activity for which a thing exists or is used. An objective is a goal. According to Locke and Latham, goals affect individual performance through four mechanisms. First, goals direct action and effort toward goal-related activities and away from unrelated activities. Second, goals energize employees. Challenging goals lead to higher employee effort than easy goals. Third, goals affect persistence. Employees exert more effort to achieve high goals. Fourth, goals motivate employees to use their existing knowledge to attain a goal or to acquire the knowledge needed to do so.[1]

Function is the activity for which a thing exists. An objective is a goal. We know from previous principles that the mission (comprised of objectives or goals) of innate intelligence is to maintain the material of the body of "living thing" in active organization (pri. 21). We also know that ALL E/Matter, including "living things," is maintained in existence by universal intelligence (pri. 1). We know that the function of E/Matter is to express instructive information (forces) created by intelligence (pri. 13) which includes living E/Matter. We also know that universal forces and E/Matter of the body must be adapted by innate intelligence for use in the body (pri. 23). We know that the objective of E/Matter, including living E/Matter, is to manifest force (instructive information) as motion (pri. 14). We also know that the function of the E/Matter of the body is to express forces (instructive information) of innate intelligence (pri. 13, 20 and 21).

Now, since force (instructive information) is manifested as motion in ALL E/Matter (pri. 14) including living E/Matter, the objective of living E/Matter is to fulfill that function by manifesting that motion with harmonious action of all its parts (pri. 32). Therefore, we conclude

1. *Read more: http://www.referenceforbusiness.com/management/Ex-Gov/Goals-and-Goal-Setting.html#ixzz3Cq9nrfNZ (http://goo.gl/Mjur5d)*

that in order for living E/Matter to be able to express innate forces (instructive information coded by innate intelligence) (pri. 13 and 20), the living E/Matter of the body is DEPENDENT on innate intelligence (pri. 20) to remain alive (pri. 21), so as to manifest innate forces which are instructive information from innate intelligence, (pri. 14 and 23) through all its parts with harmonious actions (pri. 32).

Therefore, since objectives affect individual functions, it has to be that the demands resulting from the objective of living E/Matter, (which is to be maintained alive and have all of the parts of its body functioning harmoniously with one another,) must be supplied with innate forces created by innate intelligence, thus fulfilling the principle of coordination (pri. 32).

Innate intelligence adapts universal forces into sets of instructive information that allow the E/Matter of the various components of the body to be adapted for use in the body. Innate intelligence then uses the physical brain to distribute the data through the network of nerves for coordination of action. For example, when a clerk at a bank uses a computer that clerk is using an application that understands what to do with your bank records. Applications are usually part of the hardware media, like CD-ROMs, USB flash drives and hard disks. When the clerk first starts the computer, applications are loaded into different components of the computer hardware, like processors. Once an application begins to execute its function on the hardware, it becomes a process requiring time (pri. 6). In other words, an application is a file of instructive information data ready for use, and when in use, it is a process of that file in action performing certain tasks. It uses the operating system to coordinate the activities of those tasks. In the body, this operating system is the innate brain, which is metaphysical, under the control of innate intelligence. Instructive information from innate intelligence, within the innate brain, then interfaces with the physical brain, as control center, and is transmitted through or over the nerve system (pri. 28) in order to be expressed by all the parts of the body. Thus principle #33 states precisely the law of demand and supply: "The Law of Demand and Supply is existent in the body in its ideal state; wherein the "clearing house," is the brain, innate the virtuous "banker," brain cells "clerks," and nerve cells "messengers.""

The Chiropractic Objective

Thirty-three principles comprise the foundation of chiropractic's basic science becoming the AUTHORITY of chiropractic. From those 33 Principles we can conclude that the objective of chiropractic is to locate, analyze and correct vertebral subluxation for a full expression of the innate forces of the innate intelligence of the body. PERIOD.

We note that there is NO mention of health, illness, symptoms, pain or human potential within the 33 Principles of chiropractic's basic science. For this reason it is concluded that chiropractic is NON-THERAPEUTIC and is not a part of the human potentials movement. The purpose of chiropractic is its objective, which can ONLY be deduced from its science. When a profession has its science, this very science becomes the driving FORCE (guiding instructive information) that can be disseminated to the people of the world. It is the birthright of EVERY human being to know the basic truth of the chiropractic objective. It follows that chiropractically relevant research should therefore be NON-THERAPEUTIC as well. Thus, any research directed toward a greater understanding of vertebral subluxation, its character, means and methods of detecting and correcting vertebral subluxation are chiropractically relevant. The validation of symptoms, diseases and syndromes alleviations or eliminations is NOT chiropractically relevant.

This begs the question, why is the chiropractic objective: *The location, analysis and correction of vertebral subluxation for a full expression of the innate forces of the innate intelligence of the body?* And why it is nothing else? EVERY word of the definition of the chiropractic objective is contained within 31 of those principles. Principles #1, 2, 3, 4, 5, 7, 8, 9, 10, 11, 12, 13, 14, 15, 16, 18, 19, 20, 21, 22, 23, 24, 25, 26, 27, 28, 29, 30, 31, 32 and 33 mentioned one or more of the following words: intelligence, innate intelligence, force, body, expression, 100% (full) and subluxation. The chiropractic objective is based and articulated from 31 ABSOLUTE principles forming chiropractic's basic science. It also relates to principle #6 via principle #14 since MOTION requires TIME to be manifested within the physical realm by traveling distance across E/Matter. It also relates to principle #17 via principle #30, and since vertebral subluxations CAUSE incoordination of DIS-EASE.

Conclusion

The universal principle, that we call the major premise, is APPROPRIATED by chiropractors as the chiropractic meaning of life (existence). This is our 1+1=2 and it is ABSOLUTE, DUPLICABLE and CONSTANT. From this place we can use reasoning, mostly deductive reasoning, to elaborate on the outcomes, meanings, and manifestations of this one basic truth. It is when principle #2 is deduced from the first principle that chiropractic APPROPRIATES the major premise as its start point.

The "discovery" of a universal principle is NEW knowledge and information, which is the concern of basic science. Therefore, it is observed that the universal principle, called the major premise, is absolute, duplicable and constant. The major premise constructs the foundational platform of chiropractic's basic science. This information can be applied from this platform to deduce the other thirty-two principles of chiropractic's basic science and apply them to guide the way we practice and do our research.

According to Professor Janus Hans of Wake Forest University in Salem, N.C., *"rational authority is based on reason pertaining to the relationship between aesthetic and truth."* Therefore, since according to Webster-Miriam, a principle is a truth, it becomes crystal clear that the 33 Principles comprise the foundation of chiropractic's basic science and from them it can be concluded, through rational logic and deductive reasoning, that the 33 Principles of chiropractic's basic science are the AUTHORITY of chiropractic.

I repeat, THE 33 PRINCIPLES OF CHIROPRACTIC'S BASIC SCIENCE ARE THE AUTHORITY OF CHIROPRACTIC!

Once again, it is concluded through rational logic and deductive reasoning, based on the AUTHORITY OF THE 33 Principles of chiropractic's basic science, that the chiropractic objective is the: The location, analysis and correction of vertebral subluxation (LACVS) for a full expression of the innate forces of the innate intelligence of the body. PERIOD.

You probably realized and concluded that chiropractic's applied science consists of practicing the chiropractic's objective, which is to LACVS for a full expression of the innate FORCES of the innate intelligence of the body and you are absolutely correct! Remember that at the beginning of this book, we explained that according to Seton Hall University's Department of Science and Technology Center, it is precisely "Basic science that is concerned with the process of discovery. Basic scientists seek to discover new knowledge and information without the primary concern of how the principles they create might be used. Applied science takes information that already exists and utilizes it for the solution of an existing problem." Here we have the chiropractic objective, which is the direct conclusion of the 33 Principles of chiropractic's basic science. The problem that needs to be solved is interference with innate FORCES (pri. 29) CAUSED by vertebral subluxations (pri. 31). The solution to correct vertebral subluxations is to APPLY an intelligent external universal force in the form of a specific adjustic thrust with the intent and expectation that the innate intelligence of the body will perform a vertebral adjustment, thereby correcting the vertebral subluxation. Chiropractic, like all other scientific discipline (physics, chemistry, biology, etc.), has basic and applied aspects. Again, basic science is more basic in the sense that without discovery of PRINCIPLES (emphasis mine), there is nothing to apply. Applied science relies on and could not exist without basic science.

Stated in a different way, you probably realized and concluded by now that chiropractic's applied science is derived from the 33 Principles of chiropractic's science and consists of practicing the chiropractic objective. And if you did, you are absolutely correct in your conclusion. Remember that at the beginning of this book, we mentioned that according to Seton Hall University's Department of Science and Technology Center, it is precisely "basic science that is concerned with the process of discovery. Basic scientists seek to discover new knowledge and information without the primary concern of how the principles they create might be used. Applied science takes information that already exists and utilizes it for the solution of an existing problem. The problem that was uncovered by D.D. Palmer on September 18, 1895 was the discovery of the limitation of adaptation, which is furthered by vertebral subluxation, which is a CAUSE (pri. 31) of interference to the expression of the innate FORCES of the innate intelligence of the body (pri. 29). Chiropractic was born from the discovery of a NEW knowledge and information that gave rise to 33 Principles creating its basic science. The conclusion of the 33 Principles of chiropractic's basic science is the chiropractic objective, which is to locate, analyze and facilitate the correction of vertebral subluxations (LACVS) for a full expression of the innate FORCES of the innate intelligence of the body... period. The 33 Principles of chiropractic's basic science form a solid foundation from which to provide the solution to the existing problem of the interference of innate FORCES (pri. 29) CAUSED by vertebral subluxations (pri. 31). This solution to this problem, is the application of the 33 Principles of chiropractic's basic science, which becomes chiropractic's applied science, which is to practice the chiropractic objective. Again, Seton Hall University did mention earlier that, *"all scientific disciplines (physics, chemistry, biology, etc.) have basic and applied aspects. Basic science is more basic in the sense that without discovery of PRINCIPLES* (emphasis mine)*, there is nothing to apply. Applied science relies on and could not exist without basic science."*

For example, as we mentioned earlier in this book, aviation uses two basic laws, the law of gravity and the law of aerodynamic (lift, drag and thrust) and applies it to its objective which is flying. In order for an aircraft to fly, the problem for the Wright brothers was the law of gravity. They had to discover NEW universal principles of physics to create a NEW basic science... and they did! They formulated the basic science of aerodynamic. There are three basic influences to be considered in aerodynamics: thrust, which moves an airplane forward; drag, which pulls it back; and lift, which keeps it airborne. Lift comes from three theories: Bernoulli's principle, the Coandã effect, and Newton's third law of motion. With this NEW principle of aerodynamic, Orville and Wilbur solved their problem. The rest is history.

As you can see, all universal principles work together to create many different applications for many uses. In the same way, chiropractic uses three basic laws, the law of organization (universal intelligence), the limits of adaptation, and the law of ACTIVE organization (innate intelligence). The problem for D.D., B.J., and RWS was "The Limits of Adaptation" (pri. 24) furthered by vertebral subluxations. They had to discover NEW universal principles to create a NEW basic science… and they did! They created chiropractic's basic science. The 33 Principles of chiropractic's basic science reveal very clearly a conclusion, which is the chiropractic objective. There are three basic influences to be considered in chiropractic: the function of intelligence, the function of force (instructive information) and the function of E/Matter. By applying the chiropractic objective, which is to locate, analyze and facilitate the correction of vertebral subluxations (LACVS) for a full expression of the innate FORCES of the innate intelligence of the body, PERIOD, which is the conclusion of the 33 Principles of chiropractic's basic science, the Palmers and Stephenson solved their problem. The rest is history.

It is crystal clear that the only goal of basic science is to increase the knowledge base of a particular field of study. The 33 Principles of chiropractic's basic science do just that by revealing the chiropractic objective, which is to locate, analyze and facilitate the correction of vertebral subluxations (LACVS) for a full expression of the innate FORCES of the innate intelligence of the body, PERIOD. Applied science uses the knowledge base supplied by basic science to devise solutions to specific problems. Therefore, by practicing the chiropractic objective, chiropractors solve the problem of interference with innate FORCES (pri. 29) CAUSED by vertebral subluxations (pri. 31). Therefore, chiropractic's applied science consists of the practice of the chiropractic objective. It is for this reason that chiropractic can be applied to EVERYONE regardless of creed, culture, politics, race, gender, age, or health status. It is also for this reason that chiropractic is SEPARATE and DISTINCT from EVERYTHING ELSE, since the ONLY field of study that is solely dedicate to the location, analysis and facilitation of the correction of vertebral subluxations LACVS (pri. 31) for a full expression of the innate FORCES (pri. 29) of the innate intelligence of the body is: CHIROPRACTIC.

It is further concluded that chiropractic is SEPARATE and DISTINCT from EVERYTHING ELSE and is INCLUSIVE of EVERYONE regardless of creed, culture, politics, race, gender, age, or health status.

An objective chiropractor is one who accepts that the start point of chiropractic is its major premise and the end point of chiropractic is its objective. An objective chiropractor is ALSO one who chooses to apply the 33 Principles of chiropractic's basic science by practicing ONLY the chiropractic objective. Period.

The abstract concepts of the principles of chiropractic contain an air of mathematical perfection. Concepts like universal intelligence, innate intelligence, universal force and innate force are really "pure" terms that make chiropractic a pure science or what is called a basic science. I must point out that, objective chiropractors work in conjunction with the law of ACTIVE organization and that is everything to them, and it is everything, they might say. That may sound like an overstatement or mere poetry, but objective chiropractors are not rebels against anything except vertebral subluxations interfering with the expression of the innate forces of innate intelligence within themselves or others. Unfortunately many chiropractors leave the public unschooled in this regard by not sharing the chiropractic objective with them.

Objective chiropractors, those who have chosen to practice the chiropractic objective and work in conjunction with the law of active organization, are focused on life and life for all. If they are not focused, they are not true to themselves. Objective chiropractors have a great mission to perform. Not to improve the basic law, this is impossible, but to remove one of the most negative obstructions brought about by the most tragic consequence of the limits of adaptation, to further a fuller expression of what the law of active organization demands in every stage of living. Chiropractic is about life!

If objective chiropractors would band together to do this, chiropractic could be brought nearer to its capacity to serve the world as a whole. If chiropractic were given a chance to perform its objective, it could do much to further the expression of instructive information from innate intelligence for ALL vertebrata. The time is NOW for chiropractors to choose to practice the chiropractic objective under the AUTHORITY of chiropractic's basic science… with its 33 Principles.

THERE IS NOW SO MUCH FOR ALL TO GAIN!

Appendix

There are thirty-three principles that sum up chiropractic's basic science. They were articulated by R. W. Stephenson, DC, PhC in 1927. I share them here to give you a better understanding of chiropractic's basic science.

Principle One: The Major Premise.

A Universal Intelligence is in all matter and continually gives to it all its properties and actions, thus maintaining it in existence.

Principle Two: The Chiropractic Meaning of Life.

The expression of this intelligence through matter is the Chiropractic meaning of life.

Principle Three: The Union of Intelligence and Matter.

Life is necessarily the union of intelligence and matter.

Principle Four: The Triune of Life.

Life is a triunity having three necessary united factors, namely, Intelligence, Force and Matter.

Principle Five: The Perfection of the Triune.

In order to have 100% Life, there must be 100% Intelligence, 100% Force, 100% Matter.

Principle Six: The Principle of Time.

There is no process that does not require time.

Principle Seven: The Amount of Intelligence in Matter.

The amount of intelligence for any given amount of matter is 100%, and is always proportional to its requirements.

Principle Eight: The Function of Intelligence.

The function of intelligence is to create force.

Principle Nine: The Amount of Force Created by Intelligence.

The amount of force created by intelligence is always 100%.

Principle Ten: The Function of Force.

The function of force is to unite intelligence and matter.

Principle Eleven: The Character of Universal Forces.

The forces of Universal Intelligence are manifested by physical laws; are unswerving and unadapted, and have no solicitude for the structures in which they work.

Principle Twelve: Interference with Transmission of Universal Forces.

There can be interference with the transmission of universal forces.

Principle Thirteen: The Function of Matter.

The function of matter is to express force.

Principle Fourteen: Universal Life.

Force is manifested by motion in matter; all matter has motion, therefore there is universal life in all matter.

Principle Fifteen: No Motion without Effort of Force.

Matter can have no motion without the application of force by intelligence.

Principle Sixteen: Intelligence in both Organic and Inorganic Matter.

Universal Intelligence give force to both organic and inorganic matter.

Principle Seventeen: Cause and Effect.

Every effect has a cause and every cause has effects.

Principle Eighteen: Evidence of Life.

The signs of life are evidence of the intelligence of life.

Principle Nineteen: Organic Matter.

The material of the body of a "living thing" is organized matter.

Principle Twenty: Innate Intelligence.

A "living thing" has an inborn intelligence within its body, called Innate Intelligence.

Principle Twenty-One: The Mission of Innate Intelligence.

The mission of Innate Intelligence is to maintain the material of the body of a "living thing" in active organization.

Principle Twenty-Two: The Amount of Innate Intelligence.

There is 100% of Innate Intelligence in every "living thing," the requisite amount, proportional to its organization.

Principle Twenty-Three: The Function of Innate Intelligence.

The function of Innate Intelligence is to adapt universal forces and matter for use in the body, so that all parts of the body will have co-ordinated action for mutual benefit.

Principle Twenty-Four: The Limits of Adaptation.
Innate Intelligence adapts forces and matter for the body as long as it can do so without breaking a universal law, or Innate Intelligence is limited to the limitations of matter.

Principle Twenty-Five: The Character of Innate Forces.
The forces of Innate Intelligence never injure or destroy the structures in which they work.

Principle Twenty-Six: Comparison of Universal and Innate Forces.
In order to carry on the universal cycle of life, Universal forces are destructive, and Innate forces constructive, as regards to structural matter.

Principle Twenty-Seven: The Normality of Innate Intelligence.
Innate Intelligence is always normal and its function is always normal.

Principle Twenty-Eight: The Conductors of Innate Forces.
The forces of Innate Intelligence operate through or over the nervous system in animal bodies.

Principle Twenty-Nine: Interference with Transmission of Innate Forces.
There can be interference with the transmission of Innate forces.

Principle Thirty: The Causes of Dis-ease.
Interference with the transmission of Innate forces causes inco-ordination of dis-ease.

Principle Thirty-One: Subluxations.
Interference with transmission in the body is always directly or indirectly due to subluxations in the spinal column.

Principle Thirty-Two: The Principle of Coordination.

Coordination is the principle of harmonious action of all parts of an organism, in fulfilling their offices and purposes.

Principle Thirty-Three: The Law of Demand and Supply.

The Law of Demand and Supply is existent in the body in its ideal state; wherein the "clearing house," is the brain, Innate the virtuous "banker," brain cells "clerks," and nerve cells "messengers.

Final Conclusion

The final conclusion of the 33 Principles of chiropractic's basic science is the chiropractic objective.

Using rational logic and deductive reasoning, the 33 Principles of chiropractic's basic science, reveal the chiropractic objective as their conclusion. The chiropractic objective is the location, analysis and facilitation of the correction of vertebral subluxation for a full expression of the innate FORCES of the innate intelligence of the body. PERIOD. Therefore, the practice of the chiropractic objective is chiropractic's applied science, which becomes the art of chiropractic.

GLOSSARY
(CHIROPRACTIC'S UNIQUE LEXICON)

In order to continue our exploration of looking at the OLD centrally core process of chiropractic in a NEW way, we must be on the same page. Moving forward and evolving consciously together, without condemnation, requires that we agree on terms. I have compiled, with the help of Joe Strauss, a glossary of terms directly from the Chiropractic Textbook by RW Stephenson for use in this book. I have also added some NEW terms that I believe are uniquely needed for practicing the chiropractic objective.

1. **Adaptability (sign of life):** The intellectual ability that an organism possesses of responding to all information (forces) which come to it, whether innate or universal.

2. **Intellectual adaptation:** The mental process of innate intelligence to plan ways and means of using or circumventing universal information (forces).

3. **Adaptation:** The movement of an organism or any of its parts; or the structural change in that organism, to use or to circumvent environmental information (forces). Adaptation is a continuous process — continually varying, it is never constant and unvarying as are other universal laws. Adaptation is a universal principle — the only one of its kind. It is the principle of change, and the changes are always according to law, which is intellectual adaptation.

4. **Assimilation (sign of life):** The power of assimilation is the ability of an organism to take into its body food materials selectively, and make them a part of itself according to a system or intelligent plan.

5. **Innate brain:** a) That part of the brain used by innate intelligence (law of ACTIVE organization), as an organ, in which to assemble mental impulses. b) That part of the brain used by innate intelligence (law

of ACTIVE organization), as an organ, in which to adapt universal information (forces) and assemble them into foruns.

6. **Educated brain:** That part of the brain used by innate intelligence (law of ACTIVE organization), as an organ, for reason, memory, education, and the so-called voluntary functions.

7. Physical brain: That part of the brain used by innate intelligence (law of ACTIVE organization) , as an organ, to transmit mental impulses for coordination of activities of all the parts of a living organism.

8. **Educated Intelligence:** The capability of the educated brain to function. It starts at 0% at birth and evolved to 100% at death.

9. **E/Matter:** This term means energy/matter. Since $E=mc^2$, energy and matter are interchangeable since energy is simply a different configuration (properties) of electrons, protons and neutrons with varying velocities (activities). Ex: water has 2 molecules of hydrogen and 1 molecule of oxygen, whether it is in a fluid state, ice state, or vapor state. It is dependent upon the movement of its basic elements.

10. **Disease and DIS-EASE:** Disease is a term used by physicians for sickness. To them it is an entity and is worthy of a name, hence diagnosis. DIS-EASE is a chiropractic term meaning not having ease; or lack of ease. It is lack of entity. It is a condition of E/Matter when it does not have the property of ease. Ease is the entity, and DIS-EASE the lack of it.

11. **Educated mind:** Educated mind is the activity of innate intelligence (law of ACTIVE organization) in the educated brain as an organ. The product of this activity is educated thoughts; such as, reasoning, will, memory, etc. Innate intelligence (law of ACTIVE organization) controls the functions of the "voluntary" organs via the educated brain. Educated thoughts are mostly for adaptation to things external to the body.

12. Mental forces: A mental force is that something which is information uniting intelligence and E/Matter. It is transmitted by nerves for coordination of activities and is called mental impulse because it impels tissue cells to intelligent action.

13. **Universal forces:** Universal forces are universal information created by universal intelligence (law of organization) which are subjected to physical laws, and are not adapted for structural constructive purposes.

14. **Invasive forces:** Invasive forces are universal information which act powerfully upon tissue in spite of the innate resistance of the body; or in case the resistance is lowered.

15. **Penetrative forces:** Penetrative forces are invasive forces; they are information which act powerfully assailing the body and that have effect upon tissue, in spite of the innate resistance of the body;

16. **Innate forces:** Innate forces are universal forces that are adapted by the law of ACTIVE organization and arranged for use in the body. They are universal forces assembled or adapted for dynamic functional power to cause tissue cells to function; or to offer resistance to the environment.

17. **Resistive forces:** Resistive forces are internal innate information opposing invasive or penetrative forces. They may be in many forms... as physical, chemical, or mechanical. They are not called resistive forces unless they are of that character.

18. **Growth (sign of life):** The power of growth is the ability to expand according to intelligent plan to mature in size, and is dependent upon the power of assimilation.

19. **Impressions:** The information of the tissue cell to innate intelligence (the law of ACTIVE organization) concerning its welfare and doings.

20. Innate mind: The activity of innate intelligence (law of ACTIVE organization) in the innate brain as an organ.

21. **Mental impulse:** A unit of information for a specific tissue cell, for a specific function. Specific information to a tissue cell for the present moment.

22. **Poison:** Poison is any substance introduced into or manufactured within the living body which the law of ACTIVE organization (innate intelligence) cannot use in metabolism.

23. **The chiropractic definition of vertebral subluxation:** A vertebral subluxation is a condition of a vertebra that has lost its proper juxtaposition with the one above or the one below, or both... to an extent less than a lunation... which impinges upon a nerve and interferes with the transmission of mental impulses.

24. **Vibration:** The motion of a tissue cell in performing its function.

25. **Objective of chiropractic:** The chiropractic objective is to locate, analyze and facilitate the correction of vertebral subluxations for

the full expression of the innate forces (information) of the innate intelligence of the body. PERIOD!

26. **Educated universal information (forces):** Educated universal forces (information) are forces used by people for so-called voluntary functions with limited intelligent direction.

27. **Vertebral adjustment:** A vertebral adjustment is a universal force (information) adapted by the law of ACTIVE organization (innate intelligence) for the correction of a vertebral subluxation.

28. **Adjustic thrust:** An adjusted thrust is a specific educated universal force (information) introduced into a subluxated vertebra of a person by a chiropractor with the intent that the law of ACTIVE organization (innate intelligence) will produce a vertebral adjustment.

29. **Objective chiropractor (OC):** An objective chiropractor (OC) is a chiropractor WHO chooses to practice ONLY the chiropractic objective. Also called an objective straight chiropractor (OSC) and a non-therapeutic objective straight chiropractor (NTOSC).

30. **Matter:** Electrons, protons, and neutrons configured at less than the square of the speed of light.

31. **Energy:** Electrons, protons, and neutrons configured at the square of the speed of light.

32. **E/Matter:** Term reminding us that energy and matter are interchangeable as per $E=mc^2$

33. **Information:** Coded instruction to configure electrons, protons, neutrons, and their velocities.

34. **Infocosmic:** Of the Triune of Universal Life.

Intelligence (Law of Organization) - Force (Information) - Matter (Energy/Matter)

About The Author

Claude Lessard, D.C. graduated from Sherman College of Chiropractic in 1977 where he received both, The B.J. Palmer Philosophy Distinction Award and The B.J. Palmer Clinical Excellence Award. Dr. Lessard was a Charter Founder and professor at ADIO Institute of Straight Chiropractic (later became Pennsylvania College of Straight Chiropractic) where he taught from 1978 to 1980 and where became ADIO's first Director of its Community Health Center. Dr. Lessard has written an other book for laypeople titled *"Chiropractic... Amazing, isn't it?"* in 2003, which has been translated in Spanish in 2010, French in 2016 and is currently being translated in Japanese. In 1992, Dr. Lessard was elected Chiropractor of the Year by Markson's Management Services, in Long Island, N.Y.; in 1993, he was elected Chiropractor of the Year by Quest Management Systems, in Philadelphia, Pa.; and in 2006, he was elected Chiropractor of the Year by Sherman College of Straight Chiropractic, in Spartanburg, S.C.

He has lectured throughout the United States and Canada on the subject of objective chiropractic. He resides with his wife, Sara, in Yardley, Pennsylvania, where he has maintained a private practice for forty years.

Claude Lessard, D.C.

B.S. Limestone College, Gaffney, S.C.	1977
Doctor of Chiropractic Degree, Sherman College of Straight Chiropractic (S.C.S.C), Spartanburg, S.C.	1977
Internship, S.C.S.C.	1977
Recipient of the B.J. Palmer Chiropractic Philosophy Distinction Award, S.C.S.C.	1977
Diplomate of the National Board of Chiropractic Examiners	
Certified for Preliminary Professional Education #C35301, Commonwealth of Pennsylvania	
Commonwealth of Pennsylvania License #DC-1702-L	
Co-Founder and Charter Member of ADIO Institute of Straight Chiropractic	1978
Student Referral Counselor, ADIO I.S.C.	1978-1981
Assistant Professor of Chiropractic Philosophy, ADIO 1.S.C.	1978-1980
Administrative Dean of ADIO I.S.C.	1979-1980
Associate Professor of Chiropractic Technique, ADIO 1.S.C.	1980-1981
Director Community Health Center, ADIO 1.S.C.	1980-1981
Member Chiropractic Life Fellowship of Pennsylvania	
Member of the Federation of Straight Chiropractors Organization (F.S.C.O.)	
Graduate of Church Ministry Program, St. Charles Borromeo Seminary	1983-1987
Certified Myotech Examiner	
Chiropractor of the Month Award, Markson Management Services	1988
Chiropractor of the Year Award, Markson Management Services	1992
Post Graduate Course of Study in Applied Spinal Biomechanics From the Aragona Spinal Biomechanic Engineering Laboratory, Inc.	1992

Chiropractor of the Year Award, Quest Management Systems	1993
Member of the Distinguished Board of Regents, S.C.S.C.	Since 1993
Member of Parker Chiropractic Resources Foundation	
Chair and Co-Author of "Spirit of '76", S.C.S.C.	1996
Founder of Clients Association for Chiropractic Education (C.A.C.E.)	1997
Licensed Private Pilot, Single Engine Airplanes Land	1998
Founder of Lessard Institute for Chiropractic Clients	1998
Recipient of the Spirit of Sherman College of Straight Chiropractic Award	1999
Licensed Pilot, Instrument Airplanes	2000
Author of "Chiropractic ... Amazing Isn't It?"	2003
Chiropractor of the Year, S.C.S.C.	2006
Motion De Felicitations, Ville De Ste. Anne De Beaupre, Resolutions 5553-09-06	2006
Pulstar Examiner	2008
Translation of "Chiropractic ... Amazing Isn't It?" In French	2008
Translation of "Chiropractic ... Amazing Isn't It?" In Spanish	2009
Autor del Libro "Quiropraxia No Es Asombrosa?"	2010
Auteur du Livre "La Chiropratique, Incroyable N'est-Ce Pas?"	2012
Author of Blue Book "A New Look at Chiropractic Basic Science"	2017
Autor del Libro Azul "Una nueva mirada a la Ciencia Básica de la Quiropráctica "	2019